SURVIVING

TORNADOES

Marne Ventura and
Heather Kissock

MEDIA ENHANCED BOOKS
AV2 BY WEIGL
ADDED VALUE • AUDIO VISUAL

www.av2books.com

Published by AV^2 by Weigl
350 5th Avenue, 59th Floor
New York, NY 10118
Website: www.av2books.com

Library of Congress Cataloging-in-Publication Data
Names: Ventura, Marne, author.
Title: Tornadoes / Marne Ventura.
Description: New York, NY : AV2 by Weigl, [2019] | Series: Surviving | Audience: Grades 4 to 6. | Includes index.
Identifiers: LCCN 2018053504 (print) | LCCN 2018056710 (ebook) | ISBN 9781489697912 (Multi User ebook) | ISBN 9781489697929 (Single User ebook) | ISBN 9781489697899 (hardcover : alk. paper) | ISBN 9781489697905 (softcover : alk. paper)
Subjects: LCSH: Tornadoes--Juvenile literature. | Tornadoes--Safety measures--Juvenile literature. | Survival--Juvenile literature.
Classification: LCC QC955.2 (ebook) | LCC QC955.2 .V465 2019 (print) | DDC 613.6/9--dc23
LC record available at https://lccn.loc.gov/2018053504

Printed in the United States of America in Brainerd, Minnesota
1 2 3 4 5 6 7 8 9 0 22 21 20 19 18

122018
120518

Project Coordinator: Heather Kissock   Designer: Ana María Vidal

Every reasonable effort has been made to trace ownership and to obtain permission to reprint copyright material. The publishers would be pleased to have any errors or omissions brought to their attention so that they may be corrected in subsequent printings.

Weigl acknowledges Getty Images, Alamy, iStock, and Shutterstock as its primary image suppliers for this title.

First published by The Child's World in 2016.

TORNADOES

Contents

Chapter 1

Tornadoes can happen any time, but most happen during the day.

Twister!

Wilma Nelson's **weather radio** turned on. It crackled and woke her up. Then, it sounded a tornado warning. She grabbed her cell phone and glasses and quickly went into her closet. She called for her dog. But it ran away. Nelson had heard a twister might be on its way. She had gathered supplies in the closet before she went to bed.

Nelson heard a sound like an explosion. Her ears hurt. A piece of a wall fell in the closet. It bumped her on the head. She looked up. She could see the sky instead of the ceiling. She tried to stay calm.

As Nelson waited, a tornado swept through her hometown of Woodward, Oklahoma. It was April 15, 2012. The twister stayed on the ground for 37 minutes. It traveled 2 miles (3 kilometers) and was about 1,300 feet (400 meters) wide. The twister killed six people. It destroyed 100 homes and damaged many others.

Nelson was stuck in the closet after the twister passed. Pieces of **debris** blocked the door. Her neighbors came looking for her. She signaled to them with a flashlight. They helped her get out and found her dog safe, too.

Nelson was fortunate to survive the tornado. She is sure her weather radio saved her life.

Enhanced Fujita Scale

Meteorologists measure tornadoes on a scale of 0 to 5. The Enhanced Fujita scale is used to rate how badly a twister damages structures and vegetation. One that blows some shingles off a roof is an EF-0. The worst possible twister is an EF-5. It can destroy an entire house. Scientists use information about how bad the damage is to **estimate** the wind speed of the twister. High wind speeds create the worst damage.

Nelson lives in a region of the United States called Tornado Alley. It has more twisters than any other area. But tornadoes strike all over Earth. They often come without warning. Nelson was ready for the disaster even when she was asleep. She knew the dangers, had a plan, and moved quickly. She knew how to survive a tornado.

Following the tornado, Oklahoma Governor Mary Fallin visited Woodward to survey damage and speak to residents of the town.

The **wind** of a tornado is the **strongest** wind on Earth.

In **1931**, a tornado in Mississippi picked up an **83-ton** (75-metric ton) **train** and tossed it **80 feet** (24 m) away.

Only about **2 percent** of tornadoes are **strong enough** to be considered violent.

Chapter 2

A car was thrown on its roof after a twister hit Berthoud, Colorado, in 2015.

Learning the Risks

A tornado, or twister, is a tube of rotating wind. It forms underneath a thunderstorm and reaches to the ground. Tornadoes can spin as fast as 300 miles (483 kilometers) per hour. You cannot see wind. But you can see twisters. They pick up water drops, dust, and other objects. They are strong enough to lift cars and houses.

Tornadoes can uproot big trees, which can fall on houses, roads, and power lines.

About 1,000 twisters touch down in the United States each year. This is more than any other country. The storms occur mainly in the central and southern United States. They happen most in the spring between 3:00 p.m. and 9:00 p.m. Most twisters last less than ten minutes. They move across the ground at roughly 30 miles (48 km) per hour. Twisters usually travel less than 15 miles (24 km) on the ground.

Twisters can touch down quickly. Sometimes, there is little or no warning. A twister surprised the people in Longview, Washington, in October 2014. It ripped the roof off a building. This sent debris flying. Some pieces went through a store's window. Twisters do not happen often in Washington. But the store's employees knew what to do. They hid under desks and tables. No one was hurt.

Falling and flying objects are a big danger when a twister hits. Garage doors and car tires blew through Sweetwater, Texas, during a tornado in 1987.

There are also dangers after a twister. These storms often leave behind dangerous debris, such as broken glass, exposed nails, and splintered wood. The tornado in Sweetwater broke water pipes. It knocked down power and telephone lines. And it threw trees onto roads. This kind of damage can make it hard for people to get around, communicate, and get help. Food spoils quickly without refrigeration. And water from broken pipes is not safe to use.

Tornado Alley

The region from central Texas to South Dakota is called Tornado Alley. These states get warm, moist wind from the Gulf of Mexico in the spring. They also get dry, cold wind from Canada. Warm air usually rises. But here, the warm air from the south gets trapped under the cold air from the north. This causes the warm air to spin. The warm air finally pushes up through the cold. This is an **updraft**. It makes the cold air spin down in a **downdraft**. This makes a twister.

Timeline

All of the continents except Antarctica have reported having tornadoes. The first ever recorded tornado took place in Great Britain on October 17, 1091. It destroyed many notable landmarks, including London Bridge.

1909 In Brinkley, Arkansas, 49 people die and hundreds are injured in a tornado on March 8.

1925 The Tri-State Tornado hits Missouri, Illinois, and Indiana on March 18. It is the deadliest tornado in U.S. history, with 695 people losing their lives.

1974 A total of 148 tornadoes hit 13 U.S. states between April 3rd and 4th. The path of damage is 2,500 miles (4,000 km). Deaths total 330 people. Another 5,484 are injured.

1989 The deadliest tornado ever recorded takes place in Bangladesh on April 26. Carving a path 1 mile (1.6 km) wide and 10 miles (16 km) long, the tornado takes the lives of approximately 1,300 people. An additional 12,000 people are injured.

2013 The largest tornado in U.S. history touches down in El Reno, Oklahoma, on May 31. With a record-breaking width of 2.6 miles (4.2 km), it produces winds of almost 300 miles (484 km) per hour. Although running through rural areas, it still causes the deaths of eight people.

2017 Thirty-one tornadoes hit Louisiana, Mississippi, Alabama, and Georgia on January 2. There are more tornadoes reported in the United States on this one day than in the entire month of January 2016.

Chapter 3

After a tornado, search and rescue teams are sent to look for survivors. Tornadoes kill about 60 people each year in the United States.

Making a Plan

Tornadoes can happen quickly. So it is important to learn what to look for ahead of time. The sky might look green or black when a twister is coming. The wind might change direction. Heavy rain or hail might stop suddenly. You might see a funnel-shaped cloud reaching toward the ground. There might be a sound like thunder or a loud jet. Listen to the news if you think a twister is coming. **Officials** will send out information on the radio. A tornado *watch* means you should be ready. A twister might touch down. A tornado *warning* means one has been seen. You should get to a shelter right away if there is a warning.

Doppler Radar

Scientists use Doppler radar to find and track storms. Radars send radio waves into the sky. Waves run into things like raindrops and clouds and reflect back. The radar can tell how far away the storm is by how long it takes the radio waves to come back. Radio waves react differently to things that move toward and away from them. Doppler radars can pick up on these differences. This allows them to show the movement of storms.

Specially-built storm cellars provide the best protection from flying debris and falling trees.

People who are at home during a twister should go to the lowest level of the building. A cellar or basement is best. The next best place is a room on the lowest floor. Stay away from windows. Lie down flat. Protect your head with your arms or a pillow.

Sometimes, people are away from home when a twister hits. People who are driving cars should stop, get out, and find the lowest place to wait out the storm. They should lie down in a ditch if no other shelter is around and put their arms around their heads. Overpasses or bridges might seem like safe places to hide under. But they are not. Objects can fly into the tunnel at high speeds and hurt or trap people. Stopping a car under a bridge can also cause traffic accidents and block roads.

Homes in areas where twisters are common often have storm cellars or basements. A twister went through a neighborhood near Moore, Oklahoma, in 1999. It missed Dan Garland's home. But it damaged many homes close by. So he chose to make a storm cellar. Another twister hit the town in May 2013. The Garland family went into the shelter. They waited out the storm. Dan knew building the shelter was a smart choice. When he climbed out of the shelter after the storm, he saw that his house was gone.

Family members who are in different places will want to find one another after a twister. It is important to have a plan. Local phone lines may not work. But sometimes, long-distance calls go through. You can call someone who lives out of town on a landline. Family members should pick a person to call and memorize the number. They can leave messages for each other this way. Otherwise, cell phones might work, and texting may work better than calling. You can also pick a meeting place for after the storm.

In the United States, people have about **13 minutes** to find **safety** after a **tornado warning** has been issued.

Of the estimated **100,000 thunderstorms** in the United States each year, only **500 to 1,000** will produce tornadoes.

Most tornadoes are **weak** and last only **2 to 3 minutes**.

Chapter 4

It is important to have an emergency kit as help might not arrive for several days after a tornado.

Creating a Kit

On a normal day, you get water from the kitchen faucet. The refrigerator keeps your food fresh. You turn on the heater if you are cold. Electric lights help you see. Your family can drive you to the doctor if you are hurt.

Houses, hospitals, and grocery stores can be damaged or destroyed in a tornado. Clothing, blankets, and tools can be swept away. Water lines and power lines can snap. Debris can cover roads. It might take days or weeks for workers to repair the damage. Making an emergency kit can help you survive if a tornado turns a normal day into an emergency.

Think about what you and your family will need if caught in a twister. Use the list on the next page to put together a kit. Store your supplies in sturdy, waterproof bags or buckets with handles. Make sure everyone in your family knows where they are.

Predicting a Tornado

Scientists know what **conditions** must be present for tornadoes to happen. But even when the conditions are right, often tornadoes do not occur. This makes it hard to predict tornadoes. Scientists use computer **models** to help. Computer models use data from radar, weather balloons, satellites, and trained storm spotters to predict when and where a twister will touch down.

Keep your kit in the basement or storm cellar. You can make a second, smaller kit, too. Put water, food, and a first-aid kit in the trunk of your family car. That way, if you are not at home, you will still have what you need.

It might seem like the danger of tornadoes would make people avoid living in places where twisters are common. But many people live in Tornado Alley and other places where the storms are likely to happen. They understand the dangers. But they do not worry. They know that if one does strike, they will be prepared.

EMERGENCY KIT CHECKLIST

- ☐ 3 gallons (11 liters) of bottled water per person
- ☐ Canned and dried food
- ☐ Hand-operated can opener
- ☐ Battery-operated radio with extra batteries
- ☐ Flashlight and extra batteries
- ☐ First-aid kit
- ☐ Moist towelettes, trash bags, and ties
- ☐ Local maps
- ☐ Special medications or eyeglasses
- ☐ Baby or pet supplies
- ☐ Sleeping bags
- ☐ Set of clothing with sturdy shoes for each person
- ☐ Fire extinguisher

Tornado Trackers

While radar can indicate that a storm is capable of producing tornadoes, it cannot always detect a tornado. Communities around the world instead rely on trained storm spotters to recognize and report tornado activity.

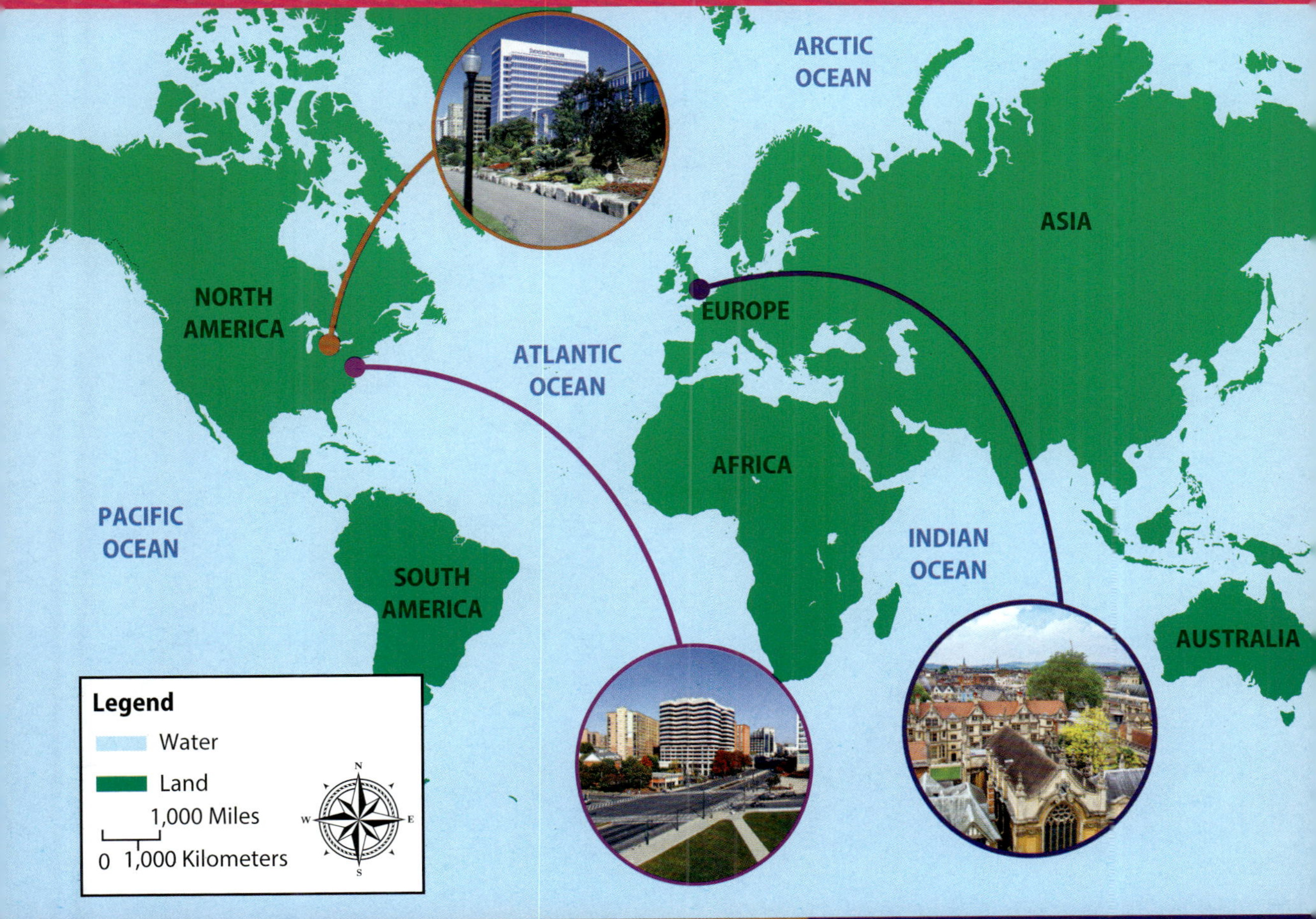

United States Based in Silver Spring, Maryland, Skywarn is a program run by the National Weather Service. It uses local storm spotters to track and report on severe weather in their areas.

Canada Established in Windsor, Ontario, Canwarn is a network of trained severe-weather spotters. They work to confirm weather conditions indicated by radar and other technologies.

Great Britain Founded in Oxford in 1974, the Tornado and Storm Research Organization (TORRO) conducts weather studies, using both storm spotters and meteorologists.

Quiz

1 What scale is used to measure tornadoes?

2 What percent of tornadoes are strong enough to be considered violent?

3 How fast can a tornado spin?

4 Approximately how any tornadoes touch down in the United States each year?

5 When do most tornadoes occur?

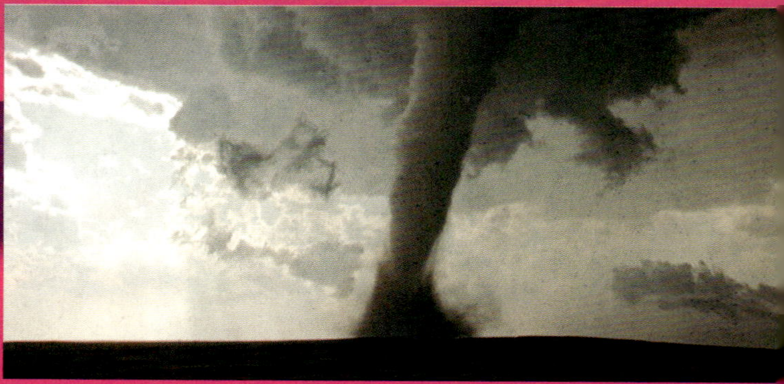

6 What is Tornado Alley?

7 Where and when did the first recorded tornado take place?

8 How many minutes do people have to find safety after a tornado warning has been issued?

NSWERS 1. Enhanced Fujita scale **2.** About 2
cent **3.** 300 miles (483 km) per hour **4.** About 1,000
Mostly in the spring between 3:00 p.m. and 9:00
n. **6.** A region stretching from central Texas to South
kota **7.** Great Britain on October 17, 1091 **8.** About
minutes

Key Words

conditions: things that must happen before something else can occur. Specific air conditions lead to tornadoes.

debris: the broken pieces of a building. Wearing sturdy shoes will protect your feet from debris.

downdraft: downward-moving air. The updraft and downdraft are both strong in a twister.

estimate: to make a smart guess about the measurement of something. Meteorologists estimate the wind speed of a tornado based on the damage it causes.

meteorologists: scientists who study weather. Meteorologists warn people when tornadoes are likely.

models: representations that show how something works. Weather models help predict twisters.

officials: people who hold public office. Officials can evacuate residents before a twister.

updraft: upward-moving air. Warm air rises, creating an updraft.

weather radio: a device able to receive emergency alerts. A weather radio can help people be ready for a twister.

Index

Log on to www.av2books.com

AV² by Weigl brings you media enhanced books that support active learning. Go to www.av2books.com, and enter the special code found on page 2 of this book. You will gain access to enriched and enhanced content that supplements and complements this book. Content includes video, audio, weblinks, quizzes, a slide show, and activities.

AV² Online Navigation

Audio
Listen to sections of the book read aloud.

Book Pages
AV² pages directly correspond to pages in the book.

Video
Watch informative video clips.

Embedded Weblinks
Gain additional information for research.

Key Words
Study vocabulary, and complete a matching word activity.

Try This!
Complete activities and hands-on experiments.

Quizzes
Test your knowledge.

Slide Show
View images and captions, and prepare a presentation.

AV² was built to bridge the gap between print and digital. We encourage you to tell us what you like and what you want to see in the future.

Sign up to be an AV² Ambassador at www.av2books.com/ambassador.

Due to the dynamic nature of the Internet, some of the URLs and activities provided as part of AV² by Weigl may have changed or ceased to exist. AV² by Weigl accepts no responsibility for any such changes. All media enhanced books are regularly monitored to update addresses and sites in a timely manner. Contact AV² by Weigl at 1-866-649-3445 or av2books@weigl.com with any questions, comments, or feedback.